# Contents

# Preface

The present mathematics team from the National Foundation for Educational Research:

**Derek Foxman**
**Graham Ruddock**
**Lynn Joffe**
**Keith Mason**

**Secretary:**
**Rajinder Jasdhoal**

**Technical assistant:**
**Rose James**

**Past team members:**

| | | | |
|---|---|---|---|
| Elizabeth Badger | (1978–81) | Peter Mitchell | (1980–83) |
| Michael Cresswell | (1977–80) | Jenny Tuson | (1977–79) |
| Richard Martini | (1979–81) | Murray Ward | (1977–78) |

# 1 Introduction

Questions about how best to measure attitudes to mathematics, what their relationship is to performance and the extent of gender differences in both attitudes and performance, are matters of ongoing discussion and research. This booklet outlines some of the findings from the APU mathematics surveys so far and is intended as a contribution to this important educational debate. This booklet is one in a series of publications in which issues of interest to mathematics teachers are highlighted.

## Why assess attitudes?

There are a number of reasons for this, including:

(1) It is important to know what pupils' thoughts and feelings are towards what they do in school.

(2) A positive approach to any school subject should be an educational goal in itself.

(3) There is increasing evidence to suggest that different groups of pupils, girls and boys for example, may be picking up different, subtle 'messages' about what is expected of them in mathematics and this may be influencing their performance in class and perhaps subsequent career choices.

To get some idea about how pupils feel about mathematics is, however, not easy. Much depends on the sorts of questions asked, how they are asked, who is asking them and what pupils see as the likely outcome of their responses.

1

## How have the APU assessed attitudes until now?

Attitudes have been assessed mainly through written questionnaires, completed by subsamples of about 1,200 pupils at ages 11 and 15 respectively, in each survey. These pupils, drawn from larger samples of about 13 000, have also taken short response written tests.

The questionnaires for both 11 and 15 year olds consist of different sections of varying degrees of structure and specificity; in one part, information is sought about mathematics in general, whilst other sections deal with specific topics and examples.

Other attitudinal data have been collected by specially trained teachers, during the practical testing interviews. In future surveys, additional information about attitudes will also be sought from pupils working together on group tasks.

## What did pupils think of the questionnaires?

B    *This is the first time I have been asked "How do I like mathematics?"*\*

For many pupils, this was the first time they had been asked how they felt about mathematics. Some pupils commented that the questionnaire had made them think about what mathematics meant to them, others expressed appreciation that their opinions had actually been sought. The questions provided them with an opportunity to reflect on situations which they normally take as given and unalterable. As a result, they appear to have given the questionnaire careful consideration, as can be seen below.

G    *I thought the questions brought out answers which I wouldn't have spoken aloud about. The paper was interesting and made you think about maths. I didn't realise I'd learned so much about maths.* (Age 11)

---

\* Throughout this booklet, indented comments in italics are taken from pupils' work, indented non-italicised statements are taken from the text of the questionnaires.

A wide range of pupils' opinions was expressed in the attitude questionnaires, not all of which was meaningfully quantifiable. However, where pupils' comments are quoted in the text of this booklet, they have been selected as being representative of sentiments that were widely expressed.

Most of the comments quoted in the text were made by 15 year olds, unless otherwise indicated. Not surprisingly, the older pupils wrote more and were better able to articulate their feelings.

B or G indicates that the comment was made by a boy or girl respectively.

B   *I think this questionnaire is very useful for scientists who are finding out about maths. I liked it very much because I have never actually realised what I did like and what I did not like.* (Age 11)

B   *I think it was good. I like the way we were allowed to say what we like about maths. If we like it or don't like it, and in my case I don't.* (Age 11)

G   *I don't think people realise how much maths they do use, it's used for more than I realised before I filled this in.*

B   *Maths should be more exciting for my age group and not so tedious. Maybe this will help. THANKS!*

## What is 'maths'?

At age 11, pupils' feelings about what constitutes 'maths' and what does not are fairly flexible. At age 15, however, pupils tend to have strong opinions about what is 'maths' and what is not. They often do not see questions involving everyday experience, for example timing a recipe, as mathematical material; in fact a number of pupils made comments to that effect

*this is not maths, it's common sense.*

During the one-to-one, practical surveys, when planning a day out, involving reading of timetables, budgeting and integrating travel, activities and meals at a fairly complex level, many pupils when asked said:

*We don't do things like this in class. It's not really maths is it; it's just . . . sort of life.*

However, when pupils find tasks more difficult, often when calculation is involved, they appear to take 'refuge' in references to 'maths'. For example, when struggling with an item on subtraction of decimals, many pupils wrote comments like

*I'm not much good at maths.*

*I know what to do, but my maths lets me down.*

In the latter case above, the pupil's method was correct, but some arithmetical errors had been made. Statements like those above are often about tasks in which the arithmetic alone is problematical for pupils, rather than the concept, but these notions appear to colour pupils' attitudes to mathematics in general; it seems that 'maths' and 'arithmetic' are seen as synonymous.

# 2 Attitudes

Collecting information about how much pupils enjoy mathematics, whether they find it useful, how easy or difficult they perceive it to be and what factors most influence the attitudes, has been a central theme of the attitude surveys.

## Is mathematics an emotive subject?

It is evident from the comments that pupils make that feelings about mathematics often run high and that it is a very emotive subject, as can be seen below. Strong, usually negative feelings are often engendered by the mere mention of the term 'maths'.

G   *I love maths.*

B   *I think that most of the sums we do in maths are absolute rubbish and an entire waste of time. You will only need basic arithmetic anyway so why waste time doing something pointless when you could be studying for another subject.*

G   *We did this type of maths once but our teacher tries to cram everything together and I don't understand. My maths does NOT deserve to be put in such a high class . . . I wish I could understand maths more clearly.*

G   *I don't like maths. I don't suppose anyone else feels the way I do.*

G   *Some boys and girls do really bad in maths. It is not because they are stupid though . . . . I myself am very bad at maths . . . you know (that people are bad at maths) by their attitude to maths . . . boys and girls panic or run out of time . . . . Just because people are no good at maths it doesn't mean to say they*

4

*are stupid. They could be really good at English. Somebody may be really terrible at English and ace at maths.*

## What about pace of work, teachers and teaching?

Comments about pace of work, teaching and teachers are common. They often indicate that pupils feel that their teachers do not recognise their difficulties. In response to a request to mention anything important about mathematics, not covered in the bulk of the questionnaire, one pupil said:

G   *Maths should be taught at a slower pace than it is because this does not give people enough time to understand it properly. And the teachers should be more understanding.*

B   *Should always be plenty of examples for boys and girls to copy down. The teacher should always explain thoroughly. Should always help if pupil doesn't get the question. The teacher should always be patient with a pupil who can't get what the teacher is trying to explain.*

Others said:

B   *People like myself would do better if we had more time to understand the work we do and then we would be able to get better marks and also have a better knowledge of maths.*

B   *I think that teachers tend to do the simple work not fully enough so that there are more problems when it comes to doing more complicated subjects.*

G   *Let the teachers explain it more easily so you know what you are doing.*

B   *...My teacher doesn't show us how to do them the easy way, but teaches us how to do it the baby way, which we find very complicated and sick, he thinks we are baby's! ...*

## What are pupils' perceptions of mathematics?

In general, do pupils enjoy mathematics, find it useful and easy or difficult?

In the more structured part of the questionnaire pupils were asked to complete three main sections. In these they were asked to express the extent of their agreement or otherwise with statements about general features of mathematics, by rating particular topics according to how difficult and useful they believed them to be. Here just the name of the topic was given. Later pupils were asked to complete representative examples of these topics and comment on them if they wished to.

## Do pupils enjoy mathematics?

When asked to state their favourite subject, about 250 out of 500 11 year olds gave mathematics as a first choice.

B   *Maths is my favourite subject.* (Age 11)

Similar information was not collected for 15 year olds, though, as might be expected, feelings about enjoyment of mathematics covered a wide range:

G   *I don't think it* (mathematics) *should be a main subject. Because some people are good at the Arts and some are good at sciences, and the ones that are not good at the sciences have to suffer – for one me.*

G   *I do not on the whole enjoy maths.*

Comments made by 11 year olds suggest that enjoyment is an important influence on their attitudes to mathematics. At age 15, pupils mention enjoyment far less frequently. They appear to take a more pragmatic approach than the younger pupils; mathematics is seen as useful and this seems to override concerns about enjoyment. This pattern of enjoyment is borne out in the structured sections of the questionnaire. What we do not know is how enjoyment, or lack of it, affects facets of mathematics not mentioned in the questionnaires.

In the structured section of the questionnaires, statements expressing opinions about the liking and enjoyment of mathematics elicit varying degrees of response, depending on the strength and direction of the opinion expressed. At age 11, for example, nearly three-quarters of pupils agree with the statement

I enjoy most things I do in maths.

However, when this statement is made more positive,

I enjoy everything I do in maths,

the distribution of responses changes dramatically; just under 40 per cent agree with this statement. So there are clearly fine gradations of opinions of this type, which are influenced by the strength of the sentiment expressed in the text.

There is a further decrease in the number of pupils agreeing, when a statement expressing a very negative opinion about mathematics is offered, like

I never feel like doing maths.

This prompted the agreement of less than 20 per cent of pupils.

When the opinions expressed are less direct, in statements such as

I'm disappointed when I miss a maths lesson,

the spread of responses is roughly equivalent (about a third) over the 'agree', 'disagree' and 'unsure' categories, respectively.

At age 15, statements reflecting enjoyment are often associated with a marked degree of ambivalence; 12 per cent of the sample are 'undecided' as can be seen from the responses below to

I enjoy working on maths problems.

The distribution changes, however, when the emphasis is altered, as in

Sometimes I work out maths problems for fun.

| AGE 15* Statements | Strongly agree | Agree | Disagree | Strongly disagree | Un- decided |
|---|---|---|---|---|---|
| I enjoy working on maths problems | 7% | 40% | 33% | 8% | 12% |
| Sometimes I work out maths problems for fun | 3% | 29% | 44% | 19% | 4% |

There is also a statistically significant sex difference in this rating, with seven per cent more girls disagreeing.

---

* This is how statements are presented in one part of the questionnaire. Pupils are asked to tick the box or the strongly agree – strongly disagree continuum, which best represents their feelings.

Responsiveness to such changes in wording point to the complexity of some feelings and indicate how carefully questions assessing attitudes need to be phrased and results interpreted.

Overall, girls at age 11 say that they enjoy mathematics slightly more than boys do. However, at age 15, the position is reversed with girls indicating that they enjoy the subject less.

## Do pupils consider mathematics to be useful?

Results from the rating scales suggest that there is a widespread recognition, 75 per cent in both age groups, that mathematics is a useful subject.

Usefulness is also mentioned frequently in comments, and pupils often relate topics and examples to their everyday lives. Not surprisingly, this happens more often with 15 year olds than 11 year olds. Pupils also often mention things that they think will be useful in the future.

G  *I think maths is an Important subject because you need to know maths in everyday situations like addition, multiplication, subtractions, Division and money (and in jobs). Even if you are not very good at maths I think it is best to try and do the things that you don't understand as best you can.*

G  *Although maths is not my favourite subject, I am glad I have learnt about it and can adapt some of the daily routine to it.*

G  *Yes this is what we ought to do more of in maths because this is going to useful in the future. If you work in a shop this type of thing could be useful. (Calculating with percentages)*

G  *I think maths is the most important subject in my school (apart from English).*

The perceived utility of mathematics is reflected in the agreement of 90 per cent of 11 year olds that

Mathematics is a very useful subject

and

Mathematics will help me to get a job one day.

Other direct expressions of usefulness produce similarly high levels of agreement.

Statements put to 15 year olds are less direct; for example,

> You won't be able to get on in life without a good knowledge of mathematics.

Fifty-nine per cent of pupils agree with this sentiment (19 per cent strongly), 32 per cent disagree (seven per cent strongly) and eight per cent are undecided.

Many pupils mentioned that they did not see the point of much of the maths they did.

G *Some maths topics should not be taught because most people will not ever again use the problems in later life. But some should be taught and the important ones should be taught very well, so every child understands. I think it is stupid having so many topics. You should just concentrate on the important ones.*

B *There should in my view be more about everyday maths in school, such as stocks and shares and calculating income tax.*

## Do pupils find mathematics easy or difficult?

In general, over 30 per cent of pupils in both age groups rate mathematics as difficult to some degree. Comments about such feelings are occasional at age 11 and frequent at age 15.

G *Divisions are quite easy unless you have to divide decimal places.* (Age 11)

B *I don't like doing problems because they are hard.* (Age 11)

G *Things like this won't sink in to my brain.* (Volume)

B *I find that under exam condition or under pressure maths is a hard subject to do. One thing I am always doing is forgetting the formula to a question. Many people I know do this.*

G *It's easy. I must be wrong.* (Sets and Venn diagrams)

Of the three factors investigated in depth in the attitude survey, the greatest ambivalence is associated with statements about difficulty, amongst 11 year olds; 25 per cent are not sure whether mathematics is difficult or not. This may be because pupils have not encountered sufficient materials to have formed strong opinions either way. Of the three factors discussed here, pupils' views about Difficulty are most strongly associated with written test performance amongst 15 year olds and are the source of the largest differences between girls' and boys' ratings. The reason why Difficulty is most strongly related to performance amongst the older pupils is a matter of conjecture – is it that 15 year olds are able to estimate their performance fairly accurately?

At age 15, ratings of Difficulty are of note; between 40–43 per cent of pupils 'agree' or 'strongly agree' with statements alluding to the difficulty of mathematics when specific aspects, like tests and problems, are mentioned:

> When it comes to doing a problem in maths, I get all the formulas mixed up.

## What about attitudes towards more specific aspects of mathematics?

In order to find out more about how pupils respond to specific aspects of mathematics, a number of topics were listed and pupils were asked to rate them on Difficulty (11 and 15 year olds) and Usefulness (15 year olds only) scales. Then pupils were asked to complete representative examples of some topics and say what they thought of them. Insights were gained into what makes pupils 'turn off' aspects of mathematics. For example, in response to a word problem about speed and velocity, one pupil said:

> *Too many words to describe the question made it seem difficult.*

Pupils often show great awareness of their strengths and weaknesses. Comments like the one below are made about several items:

B   *I am good at adding so this one is easy.* (Age 11)

G   *I know how to do these ones but I often make silly mistakes when converting to 100.*

At age 11, topics central to most curricula and primarily concerned with arithmetic, are rated as 'easy'. They also produce lower ratings of uncertainty and few pupils say they have not done such work. For example:

G   *It's easy as long as you set one number under another.* (Addition)

B   *I find subtractions one of the easiest subjects in maths.*

# Table 1: Topics rated easiest by 11 year olds

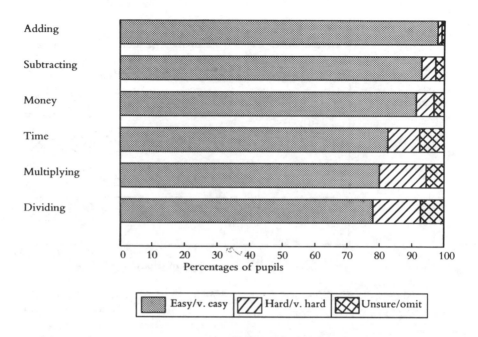

| | |
|---|---|
| Easy/v. easy | Hard/v. hard | Unsure/omit |

Topics that are considered 'hard', usually include a high proportion of 'unsure' (Table 2, p. 12) and 'not done'* (Table 3, p. 13).

G   *I can not do those sort of sums because I find angles hard to understand.*

G   *I am not very good at angles and I don't like them.*

B   *I know how to multiply fractions but am unsure as to why it works unless I think of the "×" sign as "of".*

At age 15, opinions are often strong and feelings about the difficulty and usefulness of some topics apparently firmly established.

---

* Percentages given as 'not done' indicate the proportion of pupils who ticked a box marked 'not done' in relation to a particular topic.

11

## Table 2:  Topics rated hardest by 11 year olds

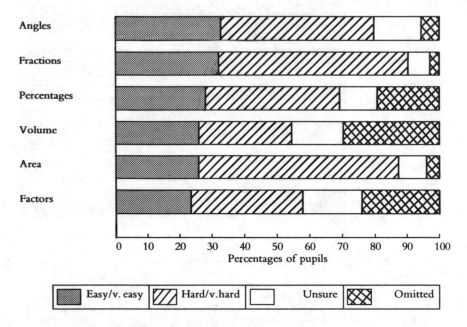

Percentages of pupils

| Easy/v. easy | Hard/v.hard | Unsure | Omitted |

The proportion of scripts left blank is between three and four per cent for ratings of Difficulty.

G   *A lot of maths problems are now made easier by calculators which we are allowed to use in school so it is easier to obtain accurate answers. I do not think that this defeats the object of maths. It makes the arithmetic easier but the theory still needs to be learnt.*

B   *I find formulas difficult as I get mixed up with the algebra.*

G   *I hate trying to figure out the formula. It is too difficult and very hard to explain.*

B   *Reading a timetable is easy because we use it in everyday life.*

As can be seen from Table 4 (page 14), those topics which are regarded as very useful have a low 'not done' rate. Topics such as trigonometry, however, are seen as being least useful, most difficult and the 'not done' rate is high.

12

## Table 3: Topics rated hardest by 15 year olds

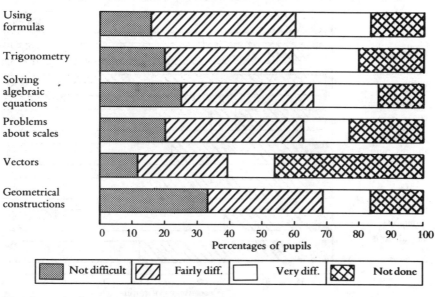

## Topics rated easiest by 15 year olds

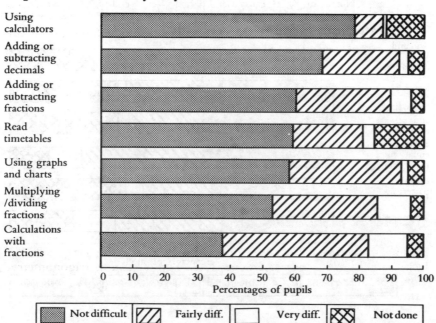

# Table 4: Topics rated most useful by 15 year olds

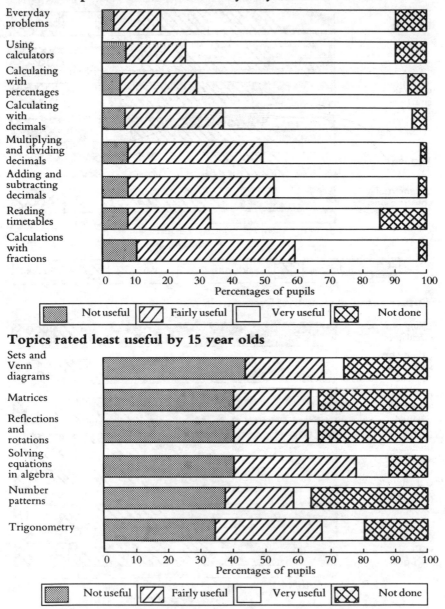

## Topics rated least useful by 15 year olds

The proportion of scripts left blank is about two per cent for ratings of Usefulness.

Some pupils elaborated on this theme:

G   *Most basic maths is useful in everyday life. These are the maths that you will need – Addition, Subtraction, Multiplication, Division etc.*

G   *Unless you are going into a specialised field e.g. work with computers, I can't see how a lot of maths we do can help, e.g. formulae, graphs, trigonometry and anything to do with triangles, and any other shape for that matter.*

B   *May come in useful for a job such as a bank clerk or wages clerk.* (Subtracting decimals)

B   *I can see no point in this. Who wants to know how to draw stupid circles with numbers in? Certainly not me!* (Venn diagrams)

G   *This question is really quite pointless. I can think of no practical use Venn diagrams can be applied to.*

## What is the relationship between pupils' ratings and written test performance?

In another section of the questionnaire, pupils have been asked to complete a written item and then rate it. An example from the questionnaire for 15 year olds is given below.

### Maths in everyday problems

To cook a joint of meat you should allow 15 minutes for each pound weight of meat and add an extra 20 minutes.

If the meat is to be cooked by 1 o'clock at what time should you start to cook a 4 lb joint?

<u>56% correct</u>

Do you find this sort of question:

| VERY EASY | EASY | DIFF-ICULT | VERY DIFF-ICULT | NOT SURE | NOT DONE/ BLANK |
|-----------|------|------------|-----------------|----------|-----------------|
| 22% | 58% | 10% | 2% | 2% | 4% |

15

Comments:

G *This sort of question shows your common sense. I can do this without a lot of difficulty. As I am a girl I need to know this kind of maths for things like cooking.*

B *It was just a ordinary take away sum gorified* [glorified?].

The results for this item illustrate a general finding for both 11 and 15 year olds, that is that pupils often underestimate the difficulty of familiar tasks; more pupils rate items as 'easy' or 'very easy' than give the correct answer.

Arithmetic items are particularly susceptible to having their difficulty underrated, as evidenced in the subtraction questions given to 11 and 15 year olds respectively. Less familiar items, like those on volume, produce more moderate ratings for both age groups, though these are still not particularly accurate, in relation to performance, especially for the younger pupils.

For 'everyday' problems, like the one about cooking a joint illustrated above, the Difficulty ratings for all pupils for the topic (when just the title 'everyday problems' is given) were compared with the ratings after such an item has been attempted. Ratings for the topic title are a more accurate reflection of success on the item than the ratings given after attempting the item. This is not the case for all items in the questionnaires.

More than 50 per cent of the sample commented on each item presented in these sections. It is interesting to note that seemingly central concepts like subtracting decimals, which might be thought of by some teachers as straightforward and unproblematical, are ones that elicit a lot of comment concerning Difficulty particularly, from all groups of pupils. In the case of subtracting decimals, this is so despite the fact that 68 per cent of pupils succeed in completing the item successfully.

An interesting finding was the frequent mismatch between performance and comments. Examination of comments reveals frequent occasions when pupils have been successful on the item but still express doubt about their performance or if they think they have completed it successfully attribute their success to luck. This appears to be more frequent in the case of girls.

G *I cannot usually do these types of sums, but I can do this one cause it's easy.*

G *I do not normally like them, but this one I do.* (Fractions, 11 year old)

G    *I always have problems with the decimal place, but not on this one it's easy.*
     (Subtracting decimals)

Some gender differences emerge from pupils' ratings. These are discussed
in the next section.

# 3 Gender Differences: Attitudes and Performance

## What gender differences are found in attitudes to mathematics?

At both 11 and 15 years old, pupils of both sexes recognise that mathematics is a useful subject. At age 15, however, boys consistently rate it as more useful than girls do.

Overall, girls at age 11 say that they enjoy mathematics slightly more than boys do. At age 15, the position is reversed with girls indicating that they enjoy the subject less.

At age 11, there is no clear pattern of Difficulty scores for boys and girls; by age 15, however, this attitude scale is the source of the most marked sex differences, with girls finding mathematics significantly more difficult statistically in all surveys.

For many other items gender differences emerge, with boys overrating the easiness of an item in relation to their success, that is, many more say it is easy than get it right. Girls tend to overrate the difficulty of more items; more girls produce correct answers for many items than say that those items are easy. Where gender differences in ratings exist, they are usually more marked in post-item rather than topic ratings. Other interesting differences between girls' and boys' approaches are discussed in a later section.

In many surveys, pupils of both ages have been asked to rate the extent of their agreement with statements about whether boys and girls perform differently in mathematics. In the 1982 survey of 15 year olds a semi-structured section, designed especially to gather more detailed information about this, was included for the first time. Pupils were asked directly (layout not as per questionnaire):

---

There has been a lot of talk recently about how well girls and boys do in maths:

Do *you* think there are any differences?    Yes/No
If *Yes*, what are they?
How do you notice them?
Where do you notice them most?
How do you think they come about?

Other comments:

---

A large proportion of both boys and girls said that there were no differences, though some of these did go on to give information to the contrary, along with those who answered 'yes'.

Amongst those who thought there were differences, opinion was divided as to which sex was superior. In favour of girls, the following comments were made:

G    *The girls tend to get better results, because they seem to be able to concentrate more.*

B    *Girls do best. Most of the boys mess around.*

B    *When questions are asked girls always seem to know the answer.*

Where differences were seen in favour of boys, the following statements were made:

B    *Most boys do maths but not many girls do maths.*

G    *In some cases I think that boys tend to have a more logical and clear view of maths.*

B    *Boys have more ambition in life therefore they work harder.*

G    *I think boys learn the more basic things and use computers etc. because they've got a better chance of getting a job to do with maths.*

Some pupils strongly refuted the suggestion of gender differences in mathematics:

B   *No, that's rubbish.*

Others were not so sure:

G   *Boys seem to take more interest in maths – but girls do quite well.*

One boy, who was ambivalent about the question of sex differences, said:

B   *It is said that boys have more ability in such subjects, I don't know what the differences are, but I do my bird's homework.*

Others said:

B   *Teachers especially men seem to pay more attention to boys in the same subjects.*

G   *Boys may need more maths to do computer studies etc. These opportunities are not given to a girl.*

G   *...when you ask a girl a maths question she takes about 5 minutes to answer it... Boys do more complicated jobs than girls, a boy learns easier as well.*

G   *Boys are more apt to use Maths in Engineering.*

G   *Both my brothers did Maths for A level which I will not do and nor did my sister.*

G   *At girls schools, people tend to go more for the Arts and people who can do Science at O level think they cannot do them for A level.*

B   *It's often thought that boys are better than girls, but I don't think so. Often girls give up when hard reasoning is involved.*

G   *If some girls do not know something, they keep quiet about it.*

G   *When a boy is asked what a question is he answers it straightaway.*

B   *The girls don't need maths as a housewife but men need them to support the house.*

B   *Boys are more intelligent than most girls.*

G   *More bosses of shops, people working in banks are men.*

## Do girls and boys respond differently to the questionnaire?

When filling in the questionnaire, boys of both ages tend to use the extreme positive ends of the rating scales ('very easy', 'not difficult' and 'very useful') far more than girls do. Girls, as a group, choose the more moderate ratings and express far more uncertainty, ticking the 'unsure' and 'undecided' categories more frequently. The extreme negative ratings are used least by both groups.

As has been mentioned briefly, for many items gender differences emerge, with boys overrating the easiness of an item in relation to their success, that is, many more say it is easy than get it right. Girls tend to overrate the difficulty of more items; more girls produce correct answers for many items than say that those items are easy. Where gender differences in ratings exist, they are usually more marked in post–item ratings rather than ratings of topic names.

For many items in the third section of the questionnaires, in which examples of topics have to be completed then rated, the proportions of girls and boys giving correct responses are very similar, if not identical. However, just looking at the correct answers does not give the full picture. When the number of incorrect responses is examined, one finds that boys also make more incorrect reponses. This results from the fact that boys actually attempt more items than girls do.*

In some written tests, multi–part items also produce dramatic gender differences in success rates; boys and girls do equally well on the first parts, but girls tend not to do as well on subsequent parts. At first it was thought that this may relate to girls' caution and unwillingness to take risks, (as evidenced by their relatively moderate ratings) since the second part often seems harder. If they are not sure, they do not answer; boys, however, tend to make some attempt.

As one boy said:

*I had a go though I don't understand.*

---

* This finding applies specifically to the items used in this section of the questionnaire which were taken from the written tests of concepts and skills. On another kind of test concerned more with problems and mathematical patterns the tendency was for boys to omit more items than the girls.

21

However, as can be seen from the multi-part items in the Problems and Patterns test, on page 26, this theory is not borne out. This leaves us with the question: what is it about the nature of the different types of items that appears to encourage girls (or not, as the case may be) to persevere?

## What are the gender differences in mathematics performance and to what extent are they a reflection of the differences in attitudes?

It has been known for some time that boys, as a group, achieve a higher level of success in public examinations in mathematics, although it should be stressed that girls' performance has been improving relative to that of boys in recent years.

Results from the APU mathematics concepts and skills surveys of pupils at all levels of attainment in their final year of compulsory schooling appear to confirm that boys' overall performance is superior to that of girls.

**It is a commonly held view that this superiority of the boys develops during secondary schooling. It is widely believed that there is no difference in performance between the sexes at age 11, or even that girls are ahead of boys at that age.**

This view of the relative performance of boys and girls at age 11 is derived from the results of standardised tests used in primary schools. These tests consist of about 50 short response items from a range of topic areas. A total score for the test is derived and the mean scores of the boys and the girls may then be compared. Typically, no differences are found, but it seems likely that the amalgamation of marks into one total score masks the differential performance of girls and boys in particular topic areas; that is, boys' superiority in some areas is balanced by girls' in others.

In the APU mathematics surveys, the success rates of boys and girls have been obtained for well over a thousand questions in both written and practical tests. This has enabled any differences in performance in different topic areas to be studied in some detail.

The main findings relate to:

i. Topic differences
   Boys and girls perform better in different areas of mathematics.

ii. Age differences
   Boys tend to be further ahead of girls at age 15 than they are at age 11, in most, but not all, areas of mathematics.

iii. Attainment band differences*
Where differences are found, the clearest differences between boys and girls are amongst high attainers (the top 10–20 per cent). The proportion of boys to girls obtaining the highest 10 per cent of scores on APU concepts and skills tests is around 3 : 2 – slightly greater at age 15 and slightly less at age 11.

Each of these is discussed in more detail on the following pages.


## (i) Topic differences

*Attitudes to topics*
At age 11, boys and girls rate most topics similarly. However, there are statistically significant differences in a few cases.

Boys rate the following topics as 'easy' more frequently:

|  | Boys (%) | Girls (%) |
| --- | --- | --- |
| Measuring | 76 | 67 |
| Length | 70 | 61 |
| Weighing | 56 | 46 |
| Area | 55 | 50 |

This opinion persists at age 15.

Girls rate the following topics as 'easy' more frequently:

|  | Boys (%) | Girls (%) |
| --- | --- | --- |
| Symmetry | 32 | 38 |
| Factors | 34 | 40 |

At age 15, girls ascribe relatively high difficulty ratings to more topics, particularly those associated with measures. Boys too rate more topics as hard, but the relative difference between the number of topics rated in this way is maintained.

It is interesting to note that, while the number of topics rated differentially by boys and girls (and which reach statistical significance) is relatively small at both ages, those dealing with Measures and Measures-related subjects are seen as consistently easier by boys and more difficult by girls

---

* Attainment bands were formed by first putting pupils in order of their overall scores on the concepts and skills tests and then placing particular proportions in groups or bands. For most APU attainment band analyses 20 per cent of pupils were placed in each band, but for some analyses the top 10 per cent were placed in a separate band.

as early as age 11. The strengths of these gender differences are more marked at age 15. The topics that boys rate as easier are also those in which they do better in both written and practical tests, as is discussed later.

*Performance on topics*
Taking a whole range of topic areas, boys tend to do best relative to girls in measures (for example on items testing knowledge of measurement units and reading scales), at both 11 and 15. Girls do best relative to boys on computations with whole numbers and decimals in either context-free exercises or in items set in money contexts.

Computations in other contexts used in APU surveys tend to favour boys. These findings are illustrated by the results for the two items below.

## (ii) Age differences

The items below also illustrate the improvement by boys relative to girls at age 15 compared with age 11 in the two topic areas which they represent: measures, and computation with whole numbers and decimals.

---

1 cm on a map represents 1 km on the ground. What is the actual distance between two towns whose distance apart on the map is 5.5 cm?

| Correct | Boys | Girls | Boys–Girls |
|---------|------|-------|-----------|
| Age 11  | 49%  | 38%   | 11%       |
| Age 15  | 87%  | 73%   | 14%       |

---

$5.07 + 1.3 =$

| Correct | Boys | Girls | Boys–Girls |
|---------|------|-------|-----------|
| Age 11  | 39%  | 51%   | −12%      |
| Age 15  | 76%  | 80%   | −4%       |

---

The actual sizes of the differences are discussed below, but essentially the actual numbers are not important at this stage – **what is significant is the differentiation in performance between measures and computation items** illustrated in the graph below.

24

Boys–Girls mean scores

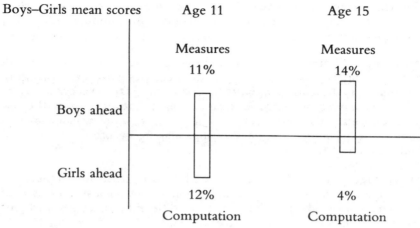

Girls–Boys mean scores

There are two comments to make about this differentiation between the measures and computation scores.

i. It is even more pronounced at age 11 than at age 15 for the two items illustrated. The differentiation in performance between the two topic areas is, therefore, present at age 11 at least as strongly as it is at age 15 – a point not generally recognised in studies where tests of heterogeneous items are used.

ii. It is much larger than the actual gains made by boys at age 15 compared with those at age 11.

In the measures item the boys' lead of 11 percentage points at age 11 is increased to 14 points at age 15 and the younger girls' 12 point lead in the computation item is diminished to four points for the older group of girls.

It is possible to conclude, therefore, that

(a) the most important differences in mathematics performance, the differentiation between measures and computation, is already present in full at age 11.

(b) the average scores of boys and girls in tests will depend on the balance of boy and girl biased items in them. Thus it is possible to make up tests which will favour either gender according to this balance.

As has been discussed earlier, this differentiation has also been detected in the attitude measures: 11 year old boys rate measurement topics such as 'area' to be easier than girls do, and girls rate some number topics to be easier than boys do.

As already indicated, the gain by boys over girls at age 15 compared with age 11 occurs in most topic areas of mathematics. **More encouraging results for girls, however, have been obtained from recent analyses of tests assessing problem solving strategies rather than mathematical content.** The processes assessed in these tests included pattern spotting, explaining rules, generalising, making up problems, and trial and error strategies.

Here, at age 11, boys' and girls' measured performance is almost the same, but at age 15, girls' scores are better overall. The tests consist of a series of written questions about each of five or six different problems or patterns. Here are some extracts from one example:

|  |  | \multicolumn{4}{c}{Success rates} |
|  |  | \multicolumn{2}{c}{Age 11} | \multicolumn{2}{c}{Age 15} |

|  |  | Boys | Girls | Boys | Girls |
|---|---|---|---|---|---|
| 1. | Complete these lines to continue the pattern. $2 + 3 = 5$ $4 + 6 = 10$ $6 + 9 = 15$ $8 + 12 = 20$ $+ =$ $+ =$ | 88% | 87% | 97% | 95% |
| 2. | This line is from the same pattern. Fill in the missing numbers. $+ 30 =$ | 54% | 53% | 83% | 86% |
| 3. | Mandy doesn't think this is from the same pattern. $+ 22 =$ Why not? Explain. | 40% | 41% | 51% | 62% |
| 4. | If you know the first number in a line of the pattern, how can you find the others? Explain. | 11% | 12% | 18% | 23% |

For the questions on this pattern the boys average success rates are half a per cent higher than girls at age 11 and two per cent below those of the girls at age 15.

### (iii) Attainment band differences

If there are more boys than girls in the top attainment band, are there more girls than boys in the bottom band? This does not appear to be the case; indeed, at age 11, there is a tendency for more boys than girls to be in the bottom 10 per cent in most topics. Thus, boys occupy more of the extreme attainment bands, while girls tend to cluster in the middle bands.

The views of pupils in different attainment bands about their own performance and about the difficulty of mathematics is illustrated by the following:

A 15 year old girl who completed most of the items correctly wrote:

G   *I find maths quite difficult, although I am a top stream candidate. However, sometimes I get really frustrated that I cannot understand my maths and sometimes feel that more time should be spent on maths in school. I feel it is one of the hardest subjects to understand.*

At the other extreme, a 15 year old boy who got few right remarked:

B   *As you may have seen, we don't do a lot of hard maths in our group, perhaps it is because it's the bottom group, but I think that we should learn the things that we did not do.*

Another pupil who was successful on over half of the items said:

B   *I think that people like myself would do better if we had more time to understand the work we do and then we would be able to get better marks and also have a better knowledge of maths.*

The sentiment in the latter remark was echoed frequently.

## Is there a relationship between performance and experience?

Differences in performance in measurement are noticeable at age 11 and by age 15, boys are doing far better. Since there are no known inherent reasons why boys should do so much better on tasks of this type, it seems reasonable to assume that some aspects of attitude, perhaps confidence, may be operating – positively in the case of boys, negatively in the case of girls. It also seems likely that if these feelings are already in operation at age 11 that they will become much stronger by age 15, unless some intervention is made.

On the face of it, it is easier to put forward an explanation of boys' gain in some areas of the mathematics curriculum during secondary schooling than it is to explain the differentiation between measures and computation in the primary school.

The notion that curricular experience may influence performance is given credence by the finding that at the secondary level there is an association between pupils' performance and their having taken mathematics-related courses. For example, pupils taking technical drawing have higher than expected scores in descriptive geometry, while those taking computer studies have higher than expected scores in algebra. Those doing wood or metal work have higher than expected scores in mensuration. Of those pupils who took mathematics-related subjects, an overwhelming majority were boys and boys tend to obtain higher scores than girls in mathematical topics related to these subjects.[*] An exception is the 1982 survey in which more girls than boys said they were taking computer studies.

Some of the largest differences in favour of boys at age 15 were found in the results of items involving visualising 3D figures. Girls do relatively much better in algebra, particularly modern algebra. Some examples of results in these areas for age 15 pupils are illustrated below.

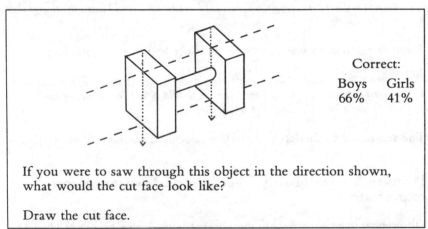

Correct:

Boys  Girls
66%   41%

If you were to saw through this object in the direction shown, what would the cut face look like?

Draw the cut face.

---

[*] In the 1982 survey technical drawing and/or wood and metal work were taken by 23 per cent of the sample. Woodwork was taken by 13 per cent and metalwork by nine per cent of the sample.

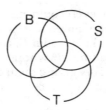

B is the set of blue shapes.
S is the set of small shapes.
T is the set of triangles.

|  | Correct: | |
|---|---|---|
|  | Boys | Girls |
| Put a tick on the space where a small, blue square would be. | 58% | 70% |
| Put a cross in the space where a large, blue triangle would be. | 56% | 68% |
| Put a △ in the space where a small, blue triangle would be. | 58% | 67% |

Fifteen year olds were asked if they use mathematics in other subjects. Most pupils mentioned Physics, Science and Chemistry. Geography, Biology and Technical Drawing were also mentioned frequently. One girl said she used mathematics in *'bookkeeping and accounts'* and said that bookkeeping was easier than maths but still difficult.

She went on:

G  *Maths is my worst subject, so, naturally I am no good at bookkeeping.*

Others commented:

G  *Maths only helps in Chemistry, whereas other subjects all intermingle in certain topics.*

B  *I find that maths in these subjects spoils my enjoyment of that subject.*

B  *Maths helps in other subjects and so is generally very useful.*

G  *You need to be quite arithmetically brained to do well at Physics 'O' level Standard.*

*G   Basic + − ÷ × crops up in all subjects.*

The results at age 11, however could not be explained by curricular differences, at least not overt ones. It is possible that boys in primary school are more likely to take a leading part in measurement activities while girls prefer doing 'sums'. At home, also, boys may be more interested than girls in activities giving them experience of measurement, rate and ratio and so on.

The understanding of decimals and fractions is important in developing pupils' abilities to read scales and therefore, to make correct and accurate measurements. It is therefore interesting to note that whereas girls at age 11 are better than boys in calculating with whole numbers and decimals, boys lead in items testing the understanding of decimal place value and have made further gains by age 15 as the results below illustrate.

| Age 11 Put these decimals in order of size, <u>smallest</u> first 0.24   0.08   0.2 | Age 15 Put these decimals in order of size, <u>smallest</u> first 0.09   0.075   0.1   0.089 |
|---|---|
| Correct<br>　　　Boys　Girls<br>　　　29%　21% | Correct<br>　　　Boys　Girls<br>　　　54%　35% |

# 4 Conclusion and Issues for Discussion

The picture of attitudes towards mathematics and the relationship to performance presented in this booklet is by no means definitive. The scope of attitudinal data that may be gleaned depends on how such information is collected. Pupils may respond differently to a written questionnaire or face-to-face interview for example. The information collected in different ways may produce different but complementary findings. The findings reported in this booklet come mainly from written questionnaires (supplemented by personal interviews) which have yielded largely consistent results over the first five-year survey period.

Since it is not in the APU brief, there has been no attempt to ascertain the effects of parents, peers and society in general on pupils' attitudes to mathematics, though other studies have given strong indications that these are an important source of influence. How the ethos of the school and individual teachers may shape or alter feelings have not been widely investigated, but is potentially a rich source of information, if pupils' comments are anything to go by.

If performance is related to attitude, then the APU findings suggest that with greater encouragement girls' performance in mathematics could be improved in relation to boys'. But boys too could benefit from greater encouragement: this report has concentrated on their performance relative to girls not their absolute performance. Also, the relative position of other groups of pupils, particularly ethnic groups, have not been monitored by the APU.

The pictures of performance seen at ages 11 and 15, with smaller differences apparent amongst the younger boys and girls, may leave junior and middle schools teachers with a false sense of complacency. On the face

31

of it, girls appear to be doing well, relative to the boys. However, there does seem to be cause for concern. How is it, for example, that even if 11 year old girls' attainment scores are no different, or even higher, than those of boys, they still express more uncertainty about their performance and ability? How is it that already at age 11 a pattern of performance is evident in which boys' scores on attitude questions are more positive and their attainment scores in measures topics higher than those of girls? Why is it that girls do better and are more confident on computational items?

It seems unlikely that the pattern of attitudes tapped in APU surveys would emerge suddenly. It is more likely that they develop throughout a pupil's schooling and therefore teachers of every age group need to be aware of the possible implications for curriculum and practice.

Finding out about attitudes requires some degree of subtlety and an investigative approach. If the questions asked are too direct, like

Do you agree that boys are better than girls at maths?

the responses are usually strongly polarised. If, however, more oblique questions are asked or other results examined (e.g. from a written or practical task), more useful data is yielded.

Pupils may also be more forthcoming if they do not know the people for whom the information is intended or if they can answer questionnaires anonymously, since they may not feel under pressure to conform or be concerned about upsetting the person who is asking. Written questionnaires, although a problem for pupils with learning difficulties, may also elicit fuller responses than oral questions which they may feel embarrassed about defending for the reasons described above. Despite these difficulties in collecting information, teachers might find it a worthwhile exercise to explore in greater depth, their pupils' feelings and opinions about the mathematics they are being taught and their methods of learning.

The following issues are offered for consideration:

How can pupils' (particularly girls') self-confidence in mathematics be increased?

How can the pace and range of work in the mathematics classroom be adapted to allow for increased understanding?

Do specific teaching approaches and/or learning modes lead to more positive attitudes to mathematics?

Is the range of experiences provided in the mathematics classroom (or elsewhere in the school) biased in favour of one group of pupils, to the possible detriment of others?

Why do specific mathematical topics seem easier to one group of pupils than another?

Are the methods of assessment used more favourable to one group than another?

You might like to ask your pupils some of the questions we have asked our national samples, things like:

— do they enjoy mathematics? Why or why not?

— do their friends share their views?

— which topics do they find difficult and why?

— how does how they feel about a mathematics lesson differ from how they feel about an English/Science/Art lesson?

Tell them that our research has found differences between how girls and boys view mathematics. Do they feel the same way? What are the differences if any?

As has been mentioned, you may find that some pupils are scared to speak frankly in front of the class, so it may be more effective to give them a few questions to answer in written form.

For further analysis of issues highlighted in this booklet, readers are referred to the following:

MASON, K. (1986). 'Areas of mathematics in which girls are ahead,' *APU Newsletter*, No. 8, Spring.

FOXMAN, D. *et al.* (1985). *A Review of Monitoring in Mathematics Peformance* (2 parts). London: Department of Education and Science.

JOFFE, L. and FOXMAN, D. (1984). 'Assessing mathematics 5. Attitudes and sex differences,' *Mathematics in School*, September.

FOXMAN, D. *et al.* (1980, 1981, 1982). *Mathematical Development Primary Reports Nos. 1, 2, 3*. London: HMSO.

FOXMAN, D. *et al.* (1980, 1981, 1982). *Mathematical Development Secondary Reports Nos. 1, 2, 3*. London: HMSO.

JOFFE, L. (1982). 'Is it your attitude that matters?' In: *Proceedings of Girls and Mathematics Association (GAMMA) Conference, Newsletter No. 4*, September.